P9-AQO-871

TOOLS FOR CAREGIVERS

- **ATOS:** 0.6
- **GRL:** C
- **WORD COUNT:** 40

- **CURRICULUM CONNECTIONS:** colors, insects, nature, patterns

Skills to Teach

- **HIGH-FREQUENCY WORDS:** a, can, has, I, is, it, like, looks, one, see, this, you
- **CONTENT WORDS:** brown, green, hair, leaf, moths, orange, red, spots
- **PUNCTUATION:** exclamation point, periods, question mark
- **WORD STUDY:** long /a/, spelled ai (hair); long /e/, spelled ea (leaf); long /e/, spelled ee (green, see); /oo/, spelled oo (looks); /ow/, spelled ow (brown)
- **TEXT TYPE:** information report

Before Reading Activities

- Read the title and give a simple statement of the main idea.
- Have students "walk" though the book and talk about what they see in the pictures.
- Introduce new vocabulary by having students predict the first letter and locate the word in the text.
- Discuss any unfamiliar concepts that are in the text.

After Reading Activities

Explain that moths are flying insects. Their wings come in many colors and patterns. These colors and patterns often provide camouflage. Explain to readers what camouflage is and why insects might need it. What was one example of wing camouflage readers saw in the book?

Tadpole Books are published by Jump!, 5357 Penn Avenue South, Minneapolis, MN 55419, www.jumplibrary.com

Copyright ©2020 Jump. International copyright reserved in all countries. No part of this book may be reproduced in any form without written permission from the publisher.

Editor: Jenna Trnka **Designer:** Michelle Sonnek

Photo Credits: rbiedermann/iStock, cover; Steven R Smith/Shutterstock, 1, 2br, 8–9; DJTaylor/Shutterstock, 2ml, 3; Sandra Standbridge/Shutterstock, 2tl, 4–5; Todd Taulman Photography/Shutterstock, 2tr, 6–7; Damian Money/Shutterstock, 2mr, 10–11; Cathy Keifer/Shutterstock, 2bl, 12–13; AlessandroZocc/Shutterstock, 14–15; Brett Hondow/Shutterstock, 16.

Library of Congress Cataloging-in-Publication Data
Names: Nilsen, Genevieve, author.
Title: I see moths / by Genevieve Nilsen.
Description: Tadpole books edition. | Minneapolis, MN: Jump!, Inc., (2020) | Series: Backyard bugs | Audience: Age 3–6. | Includes index.
Identifiers: LCCN 2018050521 (print) | LCCN 2018051538 (ebook) | ISBN 9781641288033 (ebook) | ISBN 9781641288019 (hardcover: alk. paper) | ISBN 9781641288026 (paperback)
Subjects: LCSH: Moths—Juvenile literature.
Classification: LCC QL544.2 (ebook) | LCC QL544.2 .N57 2020 (print) | DDC 595.78—dc23
LC record available at https://lccn.loc.gov/2018050521

I SEE MOTHS

by Genevieve Nilsen

TABLE OF CONTENTS

tadpole
books

WORDS TO KNOW

brown

green

hair

orange

red

spots

I SEE MOTHS

hair

I see a moth!
It has hair.

3

This one is brown.

This one is green.

spot

8

This one has spots.

This one has orange.

This one has red.

This one looks like a leaf.

Can you spot it?

LET'S REVIEW!

Moths can be many different colors. How does this moth blend in?

INDEX